CROSSING THE ABYSS
MADE EASY

CROSSING THE ABYSS
MADE EASY

Sri Clayton, Avadhuta

To order additional copies of this book, contact:
Xlibris Corporation
1-888-795-4274
www.Xlibris.com
Orders@Xlibris.com
51005

"You might say all things worth understanding are infinitely simple. And all things which are very, very difficult to understand aren't worth studying."

L.Ron Hubbard. 7 Nov 1952

PART ONE

The Abyss is said to exist on a diagram called the Quabalistic Tree of Life. This Tree was devised by Jewish mystics and taken over by Gentile occultists, such as Paul Foster Case. It is actually part of a Jewish strain of religion called Quabalah, or Kabalah, which has attracted such celebrities as Madonna. The best introduction is a book by Dion Fortune, called *the Mystical Quabalah*. The other part of Jewish mysticism taken over is numerology—each Hebrew letter has a numerical value—this is true of Greek, also, and by playing with the numerical value of words, one can squeeze out different meanings, as words with the same numereical value are considered equal to each other.

A number germane to the Thelemic tradition is 93—it is the numerical value of the word Thelema, or Will, and denotes the 93 Current, the magickal current set loose upon the world by the individuals constituting the dramatis personae of Liber AL vel Legis, the Book of the Law. Thelemites use this number in greetings and closings of letters, and also to say hello and goodbye. It is not silly, as some say, but a very lovely tradition.

The Abyss must be crossed, according to Crowleywrit, to reach the Supernals; it stands between the lower sephiroth and those which are properly part of the Godhead—namely, Binah, Chokmah, and Kether. The question arises whether transactions on the Tree are real; the authors were certainly impersonalists, and therefore lacked the highest realization. What, therefore, is Kether?

I should explain, first that my conception comes from a much older religious tradition than Judaism—from Vaishnavism, which is as old as Brahma, the religion of singing the Names of Hari, which will never accept merging into the Lord's body, and which is the yuga-dharma, or religion of the Age. Vaishnavism offers many simple answers to complicated problems, and we will find here that the problem of the Abyss is solved also.

A. What is Kether and Who is Krishna?

The first question is as to the reality of the Tree itself—it appears to be a utilitarian thing—a collection of ideas which may or may not be true—it conveys the impression of polytheism with the Ten Names of God each occupying its own sephirah, or sphere; and acting like 10 different Gods. Vaishnavism has this, also, with the 24 Vishnu forms, but in both they are all the one Vishnu, Who takes many forms.

Kether is generally considered to be the White Light—this is an aspect of Krishna—Panduranga—White Light. But Krishna is the ruler and source of the White Light:, or Brahm: "Brahmano hi pratishthaham' (Gita 14:27) "I am the source of the impersonal Brahman". Further veils of the Unmanifest simply betray the impersonal bent of the authors of the Tree. Sri Isopanisad prays: "Oh my Lord, sustainer of all that lives, your real face is covered by your dazzling effulgence. Kindly remove that covering and exhibit yourself to your pure devotee." (Mantra 15, Isopanisad).

Some consider Krishna to be a love-god, like Kamadev, or Cupid; it is true that He looks like Cupid, but His dimension is much higher; He is not a love-god, He is the Master of all Yogas (Yogeshwara), and the One without a second. Muslim mystics say that God has a very beautiful form; of course, they do not try to depict this form in any way; however, we hold that Krishna is that very form, although there are many depictions of Him. Everybody knows about Krishna and the gopis—but these are not ordinary boy-girl activities—these activities are on the highest platform of divine love. And not that many know about the chivalrous activities of Krishna as a ksatriya prince—the many battles he fought and the many kings he killed—apparently killed, for they attained liberation by being killed by Him. Krishna' activities are all non-reactive—they have no mundane result. His devotee also can live on the same non-reactive platform by understanding this—Some say there are two Krishnas—Krishna of Vrndaban, or sSyamasundar, and Krishna of Dvarka, or Vasudeva. Krishna of Vrndaban has a lot of girlfriends, and Krishna of Dvarka has 108 wives, which is actually not very many for God to have—He could have 108 billion wives and live happily with all of them by His mystic power if He so desired—it is His opulence of renunciation that He only had 108!

Krishna is seen differently by different kinds of people—very few recognize Him as the original Personality of Godhead—they see Him as their kinsman, as the most beautiful person, as Cupid, and Kamsa saw Him as Death Personified. The fact is, He does not fit easily into any "bag" as a particular type of God. Crowley thought that all Gods were partial, but Krishna is the Whole—He is not called the Supreme Personality for nothing.

This is not anthropomorphism—there was not a man who morphed into a God—Krishna is God from the beginning—He is the Form of Infinite Spirit—to say Infinite Spirit cannot take a Form is to limit it—and in fact, Form and Spirit were there from the beginning—Krishna came first—he is the Oldest One, the Ancient of Days, but He has no long grey beard—He is ever a Divine Youth—Kishor—an adolescent youth barely 16—that is His eternal Age. Krishna is sat-chid-ananda vigraha, the blissful Form of Truth and Knowledge, and His Image is the symbol of Truth, although actually his Form is inconceivable.

So where does He actually go on the Tree, apart from where Crowley put him? He goes behind the three veils of the Unmanifest. All devotees immediately agree to this, when the thing is explained to them. The Personal is higher than the Impersonal. And this, as we will see, makes travel on the Tree much easier, and will render the Abyss no more of a problem than the "water in a calves hoofprint"

B. Tiphareth—What is the Supersoul?

Yesod is simply a magnetic reflection of matters higher up on the Tree, such as the Supersoul in Tiphareth, which reflects as the "radiant form" of the Guru in the Heart. This term adequately describes the meaning and function of Paramatma, or param, super,atma, soul. Christians say, the Holy Ghost, but the Paratma is a permanent resident in all beings, even the ant, whereas the Holy Ghost only abides in Christians. It is like, there are two birds sitting on one tree; one bird is the enjoying bird, the other is the witness bird. Similarly, Paramatma witnesses and awards he fruits of actions, good or bad. He is known as the Chaitya Guru, and he takes the form of the manifest Guru to advise. He is in Tiphareth, the sphere of the Sun, and can be reached there, if you know how. Many people want to follow this chaitya guru—many varieties of "spirituality without religion" have sprung up in this way—higher self, he is known as, or higher power—the soul and the Supersoul are two—one in quality but different in quantity—it is important to know this distinction, for many equate them without differentiation—one is God, and one is Servant. Of God.

There is no spirituality without religion—religion is the nuts and bolts of the car, and spirituality is the gasoline. The great sages have recorded their conclusions, and these conclusions are the path of religion. "avibhaktam ca bhuteshu"—"Although the Supersoul appears to be divided, He is never divided. His is situated as one. Although He is the maintainer of every living entity, it is to be understood that he devours and develops all." (Gita 13:17)

One way of approaching the Supersoul is by sat-chakra yoga. He dwells in the area of the heart, on an 8-petalled lotus (some say 12 petals) as four-armed Vishnu with helmet, jewels, and the four weapons—mace, disk, lotus, and conch. With Him is Sun and Moon and lightning. It helps to say the 24 Names of Vishnu before and after this meditation. These are:

Om Sri Keshavaya swaha
Om Sri Madhavaya swaha
Om Sri Narayanaya swaha

Om Govindaya namaha
Om Vishnave namaha
Om Madhusudanaya namaha
Om Trivikramaya namaha
Om Vamanaya namaha
Om Sridharaya namaha
Om Hrshikeshaya namaha
Om Padmanabhaya namaha
Om Damodaraya namaha
Om Sankarshanaya namaha
Om Aniruddhaya namaha
Om Pradumnaya namaha
Om Purushottamaya namaha
Om Nrsimhaya namaha
Om Adhokshajaya namaha
Om Achutaya namaha
Om Janardanaya namaha
Om Upendraya namaha
Om Hariye namaha
Om Sri Krishnaya namaha

That will make everything perfect.

C. Malkuth—Who is the Guru?

Guru generally means one of the great personalities such as Chaitanya Mahaprabhu, Bhisma, Muhammed or Abraham, Christ Jesus or Srila Prabhupada. He is *as good as God*, for all purposes, because he is filled with knowledge and love of God. The Guru is never an ordinary person, and the disciple considers him to be God Himself.

The science of Guru Tattva is a very great science, and it is embodied in the understanding of the Pancha-Tattva, or God in five forms—we have heard a great deal about God in Three Forms, the Trinity—both as Father, Son, and Holy Ghost, and as Brahma, Vishnu and Siva, although the only thing these two conceptions have in common is the designation of three. We should now try to understand "Five Truths"—the Sikhs are very big on the number five—I am a bit of a Sikh myself, although I don't wear the turban and the beard and all that jazz—Sikh means the servant of the Guru—or the servant of the Guru's servants. But the modern Sikhs are simply interested in politics.

Pancha tattvatmakam Krishnam
Bhaktarupam svarupakam
Bhaktavtaram bhakitakhyam
Namami bhaktashaktikam

The Five Truths are: 1) the Lord 2)His plenary portion 3) His Incarnation 4) His energies 5) his pure devotee. The mantra goes: Sri Krishna Chaitanya, Prabhu Nityananda, Sri Advaita, Gadadhara, Srivas adi gaura bhakta vrnda. The Lord is self-evident. Nityananda Prabhu is His plenary portion—plenary means with full powers, His Incarnation is Advaita Acarya, who is an incarnation of Maha-Vishnu, from Whom all the universes come; Gadadhara represents Radharani and all the energies of Godhead, and Srivas is the pure devotee.

Advaita Acarya was the old man—that is why he is always depicted with a white beard. He was the one who originally prayed for the Lord's descent. Lord Chaitanya always treated him with the greatest respect; he did not like all this respect. So he went to a nearby town and began teaching mayavada philosophy—the philosophy the Vaishnavas hate. When the Lord heard about it, He went there and began slapping Advaita Acarya. Advaita was very pleased—"now where has all your respect gone?"

The stories of the Pancha Tattva carry us to the Supernals—"one who knows that all the associates of Lord Chaitanya are pure devotees becomes liberated." The Pancha Tattva in toto is Guru, but Nityananda Prabhu is specifically Guru, and all real gurus are His representatives.

There are many fake gurus and phony paramhamsas—this is the Age of the cheaters and the cheated—one could say nearly every "Sri" does not actually know the Supreme Truth, let alone being able to deliver it. In the ISKCON also, there are many gurus—all of them are bogus. Srila Prabhupada is the one acarya of ISKCON—he did not appoint any successors, and does not need them. They appointed themselves and got elected guru—has anybody ever heard of becoming guru by election? It is all a hoax. "A bogus guru goes to hell together with his disciples".

Lord Nityananda is the general representative of Guru—he is very powerful. He is spiritual strength personified—he could kick out the paths and toss the sephiroth like softballs. His only desire is that everyone attain Krishna prema, and he distributes it indiscriminately to sinner and saint alike.

Lord Nityananda is an avadhuta—this is a peculiar type of sadhu. The avadhuta is up to anything—he does not care very much for the strict rules, and he has very little bodily consciousness, so sometimes he appears naked.

There is a story about him—once upon a time Lord nityananda had a group of chandala disciples, and they complained to him that the rules were too hard—strict sexual abstinence, strict vegetarianism—it was too much! So by his own authority, he adjusted the rules, and allowed them a bowl of fish soup and the embraces of a young girl while chanting the Holy Names—and thus everyone became happy.

Lord Nityananda has many representatives—he is a funny guy; he is not the property of any group, or caste, or religion—he does whatever he likes. But among his current representatives are His Divine Grace A.C.Bhaktrivedanta Swami Prabhupada and my spiritual uncle, His Divine Grace B.R.Sridhar Dev Swami.

So we can accept or not—we each have our considerations in this matter, our own "disciplic succession" to consider—"One cannot be without Guru at any point in his life—to accept this guru or that is another matter".But as they told me at the Temple, many years ago, when I was coming with so many arguments and objections—"surrender is best"—put your mind aside and jump in!

There is initiating (diksha) guru and instructing (siksha) gurus—these are actually one—but for the sake of our understanding we distinguish them. The Initiator is He that connects you with the Spiritual World, the Current, or the Shabd—the Initiator is One. The instructing gurus instruct you on the many details of the meaning of spiritual/magickal life. Instructing gurus may be many—Dattatreya had 24. And Dattatreya said, "Knowledge gained from only one guru is neither very ample nor very complete."

Thus, to make up for my many offenses towards him, I look o n Jesus Christ as my initiating Guru. Actually, I was initiated by Therion directly—so whatever His realization was, I go to that. The list of my instructing gurus would be a very long list—everybody from Freidrick Neirsche to Mary Baker Eddy has helped me—but my two Vaishnava Gurus are Srila Prabhupada and Srila Bhakti Rakshak Sridhar Dev Goswami Maharaj., whose words appear in this book.

The Guru's Feet, which are placed in Malkuth and in this world, provide instant travel all over the inner and outer Universe. If we stay fast at his Feet, we can travel everywhere without leaving Them. They are the abode of all the holy places, and we can reach all these places and take bath simply by touching Guru's Feet. Certainly we can travel where we like on the Tree of Life, which is simply a meditational diagram.

There are two verses from the Guru Gita which are germane:
Om ajnana timirandasya jnananjanana shalakaya
Chaksur unmilitam yena, tasmai sri gurave namaha

He opens my darkened eyes with the torchlight of Knowledge;
I offer my obesiances to such a spiritual master.

Gurur brahma gurur vishnor gurur devo maheshwaraha
Guru sakshat Parabrahma, tasmai sri gurave namaha

The Guru is Brahma, Vishnu, and Siva
Indeed, he is the Supersoul—I offer my obesiances to such a Guru.

But the Guru is not self-created; he has his Guru, who has his Guru, and so forth, back to one of the original Personalities of the universe; our succession goes back to Brahma, at a very early time in the Universe, when science says there was nothing. There are also Vaishnava successions to the Goddess of Fortune, the Four Kumaras, and Lord Siva, at a similarly distant period.

All the acaryas of these successions are ever-living, and you may contact them also. For instance, I have a relationship with Srila Prabhupada's Guru Maharaj, Siddhanta Saraswati, who was a religious genius. He has taken an interest in me, and I admire him immensely, and Prabhupada is very pleased that I like his Guru so much.

Disciplic succession is not a constriction of thought into a narrow channel; rather, thought expands toward the Divine, takes wing, and flies. There is no limit to good, or to our access to spiritual reality.

D. The Path of the Arrow

We have spoken a great deal of Spirit and of attaining the spiritual platform—what then, of Magick? Magick is Yoga, and the Aim of the yogi is one, and he is resolute in purpose—the intelligence of those who are irresolute is many-branched. (Gita) So obviously, this Magick is not just a means to get stuff—that was accomplished in time gone by with great sacrifices like the Yajasuya—and at those sacrifices everyone was fed sumptuously and gifts were given.

So most of Thelemic practice is on the magickal plane, which includes the astral and part of the physical—it touches the spiritual planes only, but its main business is magnetic. This plane is the shadow of the spiritual. The Magician should read Science and Health with Key to the Scriptures for an understanding of Spirit and that which is not Spirit. You have doubtless heard of the old practice of neti,neti—not this, not this—better to engage in Harinam sankritan and realize directly, but as Crowley says, the East Indians are very expert at finding the most difficult way to achieve the most undesirable end!

The magickal plane is actually the kama (lust) plane—it is as mundane as a stockbroker. I have encountered mundane people who spoke very critically of magick, and there was no answering them because they and I were on the same, stockbroker plane; this was a factor that sent me very muich in the direction of becoming a transcendentalist—along with the abuse from my father.

The example is the Gnostic Mass—it is very magickal—so magickal that the spiritual is completely covered—so thoroughly that it never rises to the level of a true Mass—it is a kama tramsaction only—not that I don't like getting out on the kama plane, but Spirit rules, and I am a transcendentalist.

It should be remembered that what we think of at our last moment is what we become. So we can think of the magickal plane and become a being there—an Egyptian God or whatnot—as a daily reader of the Bhagavad Gita As It Is, I am eligible to become one of the Associates of Siva, a class of powerful demigods. I prefer, however, to go where the Guru is personally present.

The arrow flies in a straight line from Malkuth to Kether; if Kether is impersonal, this is a difficult journey; "For those whose minds are attached to the unmanifested, impersonal feature of the supreme, advancement is very troublesome." (Gita 12:5) and if one has the idea that God is also Maya, illusion, that is even more troublesome.

Actually, one who travels up the central path simultaneously gets the attainments of the surrounding paths and becomes Adeptus Major, Adeptus Exemptus, etc. without even trying for it.

"Simply by worshipping Krishna, all ones purposes are served."

The so-called Path of the Lightning Flash which the Magicians like, has it's problem—the Abyss—which does not exist in the so called "mystics approach". Actually, in devotional service we are all Magicians—transforming the world into Vaikuntha by Love under Will—that is our Great Work. And our Leader and Patron is Yogeshwara—the Master of All Mystical Powers, and our Guide is Prabhupada—the Master at whose feet many masters sit.

E. Masters of the Temple?

I have met a few self-proclaimed Masters of the Temple—they all seemed to be very angry people, not at all at peace with themselves—if that is what a Master of the Temple is, I considered I did not want to be one. It's a case, however, of anybody can claim to be anything they want (especially on the internet). So one way to cross the Abyss is to declare that one has and be done with it! We have all had a personal Abyss to cross—I remember mine very clearly; if we count those, we can all be Masters of the Temple! Alternatively, we can just lie about it, since the Abyss is a lie itself—I believe it to be simply an elaborate Crowley joke on his disciples—"there is n o Abyss anywhere in God's creation".—certainly not under the protection of the Divine Names. I should point out again: the Tree is not Ten Gods—it is One God with Ten Names. The same Diety is present everywhere on the Tree., and fills the space between the lower and Supernal sephiroth, so there is no cause for fear. The One God will place you where He wants to place you. Crowley speaks of crossing the Abyss on the strength of karma, but karma does not even reach to the supernal realm—karma is a lower pursuit—what reaches is Love and Will. What Crowley says about it is: "An immeasurable abyss divides the Supernals from all manifestations of Reason, or the human qualities of man—we find all reason identified with the abyss. Yet this abyss is the crown of the mind. It has no number, for in it all is confusion." Sounds to me like the Master had a psychotic episode on his way to Binah, and took it as a magickal fact. The Abyss is simply psychosis, and none of the great religious Traditions say that you have to go crazy to realize God. Quite the opposite—sanity gets you there.

F. Love and Will

Love and Will have a complex relationship in different situations. The question may be asked, what is Love? And what is Will? Before we go further we should have our terms defined. Crowley says Love is Change—and Change is Sorrow, according to the Buddha. We would say, rather, Love is Changeless—it is to Lust as Gold is to Iron; it is selfless, and thinks of the other, whereas Lust thinks of itself. St.John declares that "God is Love", but we consider that an incomplete, very incomplete, definition. "Love means, I enjoy or not enjoy—I love you—there is no return. Just like Radharani's love for Krishna—Krishna left Vrndaban, and their whole life remained simply crying for Krishna. Krishna never returned—but still they loved Krishna—that is Love."

"Our program is to place your love in the proper place."

(Srila Prabhupada)

And that would be by Will. So what is this Will? Apparently, everyone has one, they are all different, and they collide with great frequency, as over the color of the drapes in the living room. This is "your will" and "my will"—*human will*, based on a wrong conception of self. This will cannot take us very far, Then there is True Will, designated by the capital letter "Will"—this Will is based neither on the senses, nor the mind, nor the intelligence, nor the (false) ego which says I am this body—this Will is One, and it comes from the highest plane of soul—the platform of the higher Self.

I have had my own experience, just "touching the hem of the garment" of True Will, and what that set off in terms of so many arrangements of circumstances—no prayer could compare—it is a relationship of "simultaneous oneness and difference" with the Higher Self, in which one is one with His Will but at the same time maintains one's own identity as a man.

Jesus said, "To call me Lord is not enough; you must do the Will of my Father." The exoteric interepretation is to study the Commandments and other injunctions and deduce the Will of the Father therefrom—but better to approach that Will directly, face to Face, and hear for ourselves the Word which is His Will—chant the Holy Names! Engage in the blissful kirtan! He will speak to you in unmistakeable ways.

So where is Love in this? Love is under Will. Our love is dormant—call it original sin or conditioning, but we are antagonistic toward spiritual light and Truth, and ouor real love is covered with many psychological layers of distaste and distrust; we find people who just love the material world—nothing greater, although their circumstances are often not very good. Our hearts are very dirty, with lives after lives of accumulated errors and mistakes; we find the Holy Names to be very bitter, and taking them to be a hard job. The cure for this is to take them constantly, or regularly. Find a group of devotees who are chanting: Hare Krishna Hare Krishna Krishna Krishna Hare Hare/ Hare Rama Hare Rama Rama Rama Hare Hare, and sing with them—this will awaken the spiritual element in the heart, the soul element, and render us able to understand spiritual life.

Hara, Krishna, and Rama are the bijas (seeds) in the maha-mantra—Hara means the Goddess—Energy, and in the vocative is Hare. This mantra is received originally from Brahma, and passed down throuogh disciplic succession, where it was revitalized by Chaitanya Mahaprabhu, Who is Krishna Himself, and passed on to the 6 Goswamis, and Bhaktivinode thakur, and Siddhanta Saraswati—blessed with all their energy, it came to Srila Prabhupada, and is now appearing here.

G. Magick is Yoga

Yoga is like a ladder—one who stops at a particular step is known by that name—like, dhyana yogi, hatha yogi, etc. But the top of the ladder is bhakti yoga—the yoga of Love Divine—this attracts the divinely intellectual, who have intelligence enough to understand it and see its value. Actually, Love attracts everyone, but many cannot stay fixed in one idea. Monotheism, as Crowley says in the notes to 777, is not ordinary, so-called "normal" consciousness.

Yoginam api sarvesham—"And of all yogis, he who always abides in Me, worshipping Me in transcendental loving service, is most intimately united with Me and is the highest of all." (Gita 6:47)

So Magick is Yoga because it deals with the All High—there is the High Magick and the lower Magick—the lower Magick is getting stuff by it, whereas the High Magick is simply self-realization. Getting stuff is best accomplished by conventional means—to get women, for instance, get money, then nice car, nice clothes—spell-casting for one is black magic.

There are other Trees, and like Forms, besides the Otz Chaim—there is the Universal Form Whose feet are the nether regions and Whose head is the heavenly planets; there is the Form whose legs are the sudras, whose belly is the vaishyas, whose arms are the ksatriyas, and whose head is the Brahmins. Then there is the Tree whose "roots are up and whose branches are down; the leaves of this tree are the Vedic hymns; one who knows this tree is the knower of the Vedas."

The Magician may object to the Name of Krishna, saying that it sounds too much like Christ, and he, the magician, is an Antichristian. If you do not like the Name of Krishna, you may chant any nomenclature meant for the Supreme Being—there is room in Krishna Consciousness for Antichristians, also.

The general trouble with the Magician is that he wants to be God—he wants to be Master of everything when his constitutional position is that of Servant. Thus he takes the Names as his servants instead of engaging himself as servant of the Names. This position is insufficiently appreciated. One thinks that one is going to become a slave—but spiritual slavery is complete freedom—and there is no question of slavery due to free will in the relationship.

Question is—who is the Doer? Shastra says, the soul is the doer, but so many activities are carried on by the three modes of material nature and the soul does nothing. The modes are:

Goodness (a feeling of happiness and knowledge)
Passion (uncontrollable hankering)
Ignorance (Madness, indolence, and sleep)

We all understand what the mode of Passion is—it is very prominent in modern civilization; Magick also is very prominent—television, computers, cell phones—the mystic powers of the ancients have all been duplicated by purely materialistic means—therefore the yogis no longer crave these powers.

So there are five factors to any action, as given in Gita 18—the place, the implements, the doer, the action itself, and the Supersoul (sometimes translated as Destiny). Without the Will of the Supersoul, nothing can take place. These are the constituents of a karma.

The Magician, poor fellow, does a great deal of karma and so entangles himself further in the complexities of material nature. Three things: karma (action) akarma (non-action) and vikarma (bad action)—his activities are a combination of these things. Sometimes he thinks karma is all-in-all, but it is temporary, and can be burnt up completely by the fire of Knowledge.

I am emphasizing the trouble with Pantheism, which Crowley favored—uncertainty and unstability of Object. Unless you are convinced that God X is the Supreme Person, you cannot serve him with love and devotion. Krishna deals with this in Gita: If one wants to worship a demigod, I make his faith steady, so that he can devote himself to that Diety—surely he gets his desire, but he does not know that these gifts are awarded by Me alone."

So what is Knowledge? Krishna says, "Knowledge of the field and the knower of the field, I accept as Knowledge." There are two knowers—the soul, which knows this body, and the Supersoul, Who is also the knower, and knows all bodies. The field is nature, or the external, material energy. Nature works very wonderfully, but it cannot work without the spiritual touch. She is so powerful because She has Her backing from Krishna—therefore She is very hard to overcome. It is very easy to say, "I am not this body, I am spirit-soul" but to realize that can be the work of lifetimes. Bhagavad Gita As It Is can be a big help—it is the best manual of yoga and self-realization, spoken by the Yogeshwara Himself—it may seem trite that devotion to Krishna is Secret

Knowledge, but so many commentators on the Gita have no inkling of it. Krishna says, "There is no work that affects Me—nor do I aspire for the fruits of action. One who knows this truth about Me, O Kaunteya, also does not become entangled in the fruits of action."

So everything depends on knowledge of Krishna—our yogic/magickal careers are made or marred on this principle of life. And how to gain this knowlsdge? To know His Name is to know Krishna—the Name will tell us everything when duly approached according to discipline:

A. Daily bath.
B. Rise early—1 and a half hours before sunrise.
C. Keep the regulative principles of freedom.
  1. No illicit (casual) sex
  2. no mental speculation or gambling
  3. No meat-eating
  4. No intoxicants

D. Offer your food.

(If you break one of these principles, you will break all of them. Disciple means to follow this discipline.)

The Universe, as we hear from the Kybalion, is mental—it is Krishna's thought. We can accept the material idea, or the Divine Idea. When Adonai Elohim created the Universe, He saw that it was good. Is it good? Apparently not. But if we can accept His idea of the world and ourselves, we can become free of the illusion which keeps us here. This is a great Magick—and this is also attained simply by accepting Krishna as the Cause of all Causes and the best Friend of every living entity. This is called saranagati (surrender) and also has six steps.

So there are two Personalities to fill out the Supernals—Durga in Binah, and Siva in Chokmah. Durga rides on a tiger—she kills demons, or obstacles to material progress, and grants boons. She is also known as Parvati, the Daughter of the Mountain, and She is Lord Siva's wife. She has 1000 Names. She is white, whereas Her partial aspect, Kali Ma, is black. I do not know why so many occultists run to Kali, neglecting Durga. It may be similar to the Nrsinga people among devotees, who like the terrible man-lion Form because they do not have confidence in little Krishna with flute and morsel of food, to protect them.

Lord Siva is the Divine Ascetic. He carries a trident and resembles the Devil—He is covered with ashes, wears the three-lined tilak, and is deep in meditation upon His Lord, Rama. His companions are ghosts and goblins, and He benedicts them by placing them in the wombs of women who are not careful, thus getting them human bodies. He is very equal minded—nobody is very good, and nobody is very bad; He is also called Shambhu and Sankara and Rudra. Some say he is the Jehovah of the Bible.

So there is Siva-dhama and Durga-dhama, or Devi-dhama; Devi-dhama extends all the way down to Malkuth—Siva-dhama is it's own abode—not quite Vaikuntha. In Lord Siva's Paradise no men are allowed—if you enter, you will be transformed into a woman!

Such sex changes are found elsewhere in Vedic and Puranic literature also, both ways, showing that this transgender idea is approved by the sages.

Once I tried to make a non-religious version of Krishna's words, taking them purely as psychology. But the Gita is a profoundly religious book, and we shortly find that these statements have a religious goal—to realize the Self.

There is some discussion about the great difficulty of controlling the mind—but this topic is quickly put aside by the Speaker and never occurs again. In fact, He gives Arjuna a fairly easy practice—simply think of Him while performing one's duties; this is called Karma Yoga, which develops into Bhakti Yoga with the development of Love.

I hesitate to talk so much about "duties"—the Hindus love that word—I would rather say, "activities".because it doesn't have the connotation of something somebody with a long beard wrote down in a book that you have to do!

There are 10 Divine Names on the Tree of Life—they denote Ten Aspects, or Moods, of the One God. They are:

1. Eheieh
2. Yah
3. Jehovah Elohim
4. El
5. Elohim Gibor
6. Aloah va Daath
7. Yaweh Tzaboath
8. Elohim Tzaboath
9. Shaddai Al Chai
10. Adonai Melekh

The Jews have an aversion to saying the Names of God—this has a Biblical basis. In ancient time, only the High Priest could utter the Name of God once a year, in the Holy of Holies.

The Jehovah's Witnesses, on the other hand, have decided that Jehovah is the personal Name of God and use it for everything—but it is like a Superior Court Judge—in court, he is known as Judge or Your Honor—in his business affairs, he is Mr. Jones; to his grandkids, he is grandpa—and to his wife and friends, he is Bob—different activities, different names—same person. Same with God. According to His different activities, he has different Names. The Quabalists say that every combination of letters in the Torah is a Name of God—that is a very metaphysical and advanced view.

H. Spiritual and Mundane Gender

I believe the Kybalion calls it "polarity". Gender is not exactly the same in the spiritual worlds as it is here—"God is beyond sex—as we know it". There is a simplistic saying: Krishna is the only male—we are all females in reference to Him". This is an attempt to explain about the Energetic and the Energy in very crude terms. Actually, the Energetic and the Energy are one—like the Sun and it's light, but in Lord Chaitanya's philosophy, acintya bheda abheda tattva, some distinction is made—one and different. You will find this a very wonderful philosophy if you attempt to apply it.

So unless we live in the Supernals, we cannot know very much about spiritual sex life, although we are told there is such a thing, never to be confused with what we do here! The problem comes when advanced spiritual personalities want to hand down precepts about male and female relations in this world—then their personal bent becomes known, along with their prejudices in this regard.

In social matters, which everybody can understand, their opinions, it seems to me, are no better than anyone elses, whether or not backed by ancient texts written by men.

But thus we have very "male" religions like Islam and orthodox Judaism, and, I'm sorry to say, Krishna Consciousness. Actually, Prophet Muhammed liked women—they were his second favorite thing after Prayer. For the times, he was very progressive—he provided them with rights of inheritance and other rights they didn't have before—their situation was very much improved after the Prophet. He did not invent the burka. Krishna Consciousness, on the other hand, is so down on women that we used to refer to it as an 11th degree (homosexual) organization. This is because it has brought all the luggage of the Hindu social system with it, which is based on keeping women at the bottom of it.

Guru Nanak Dev, the first Sikh Guru, preached the equality of women at a very early time for that sort of thing. He also did not like the caste system and refused to wear a sacred thread.

In the modern, industrialized world, women do everything—they are doctors and social workers and clerks and soldiers and cops and every damn thing—I like this arrangement—they can have one man or three dozen of them and nobody cares—there is a verse in AL: "Let the woman be girt with a sword before me". The sword is the symbol of dominion—thus this means that in the new Aeon, women will have independent power. You might not know it, though, from many of the countries of the world. The old Soviet Union put an end to the oppression of women in Afghanistan, but the US sponsored the Taliban to defeat them. Now the chickens are coming home to roost.

Most all religions want to cut down the power of women—basically, they are considerably more powerful than we men, and religions don't want them loose to wreak havoc. There is "feminist religion", different kinds of wiccan groups, Starhawk and all that, and some female Christian theologians, but it is not particularly respected.

My father thought there was no value in anything "religious"—in fact he declared that anyone who became religious was doomed and would never find any good in his life. But what is the value of peace of mind? What is the value of Friendship with Krishna? What is the value of Knowledge of the difference between Spirit and matter? When one knows how to get free of the three modes and go back Home, back to Godhead, what is the practical value of that? And when one comes under the protection of a Good Guide, like Srila Prabhupada, what can one gain from the items of this world? What is the good of nothing at all—simply live like an animal and become an animal next birth—nonsense.

Human life begins with inquiry—"Athato Brahma jijnasa"—now is the time to inquire about the Greatest, the All-Encompassing. It is a question with an answer—some find it in the Qu'ran, some in the Bible, some in Liber AL, that is not so important—where they find Him is less important than the fact that He is there—some say, She is there; and this is a special order of fact which rules all other facts. I read somewhere that there is a "religion" gene—some have it, some don't—which is indicative of the two kinds of beings Krishna describes, the divine and the demoniac. The demoniac are not interested in regulating their lives for self-understanding, and that is the difference—they are not completely evil beings.

I. Every man and every woman is a star.

In modern astronomy, that means a self-luminous body. But in the old astronomy, the astronomy of the Vedas and the Quabalah, there is only one self-luminous planet in the Universe—our Sun. The stars simply reflect its light, like the Moon. This means the "stars" are not independent—the turn to the Sun for light and heat, and the Sun is the "visible Brahman"—the light of the Sun comes from Krishna's effulgence in the spiritual world. This is a physical way of showing the nature of the Vedic system of thought—everything turns to the Supreme—we are not supreme ourselves.

The Zodiac is like a clock that shows the time—similarly the arrangement of stars and planets shows the subtle influences, the planets do not "create" them, or "cause" anything any more than time is caused by the clock. It is reasonable that if the Universe is one piece of work, everything is interconnected—the devotees are accused of being superstitious because they believe in astrology—and they are! But a good Vedic astrologer can tell you about past events he couldn't possibly know by his unaided senses.

There is a star called Dhruva—better known as the Pole Star. It is a Vaikuntha planet in this Universe, and it is ruled by Maharaj Dhruva, who obtained it from the Supreme Lord as a boon. Those who are not perfect in devotional service can go there to live.

Krishna takes two forms in this Universe—Ksirodaksayi Vishnu and Garbodakshayi Vishnu. The Ksiro (Ksira means milk) daksayi Vishnu has his own planet on the northern top of the Universe—there there is a great ocean of milk where the Lord resides on the back of the Sesa (great Serpent) incarnation of Baladeva. The Universes are shaped like footballs—they have seven coverings, earth, water, fire, etc. each covering ten times greater than the last one, and they float in the Causal Ocean. To reach the Vaikuntha (spiritual) planets, one must penetrate these coverings and cross the Viraja River, which separates the spiritual and material worlds—Chiron comes to mind—then go up through the impersonal brahm effulgence to reach the Vaikunthas, where all the inhabitants are in agreement. The Vaikuntha planets are many, many times greater than the largest material Universe, and each has a predominating Form of Krishna. The inhabitants all look exactly like the Lord. The highest planet, Krishnaloka or Goloka Vrndaban, is shaped like a lotus. Grady McMurtry, the Caliph of the OTO, suggested once that it could be eaten by the Cow of Hathor—I think Krishna and the Gopis would like that!

These are just little explanations from the Vedas about how the Universe actually is situated—among those in agreement is Hermes Trismegistus—the

Universe is, actually, Krishna's thought, developed by Lord Brahma—as are we all Krishna's thoughts, and Krishna is All-Good. Therefore we are all-good also, except that we believe otherwise—tough for us! For just as He is the Complete, all beings emanating from Him are complete, also, and have everything they need to live and be happy. And yet we are engaged in a hard struggle for survival! Why/ "Allah does no injustice to anyone—but people are unjust to each other" (Qu'ran, Hud) We create our own hells here—and heavens, sometimes, too. Sometimes, the Qu'ran says, you may break open a rock and find a little frog inside—Allah provides for that frog, so He certainly will provide for us—in this is a Sign for those who can see.

Sometimes it is like Sartre says, "Hell is other people"—it is certainly that way in the jail—the problem is not the confinement—it is the people you have to be confined with! So the worlds extend downward to Patala Loka, seven stages down, which is a darkly beautiful place—it is Narada's favorite place—much further down, the Hells begin—some are cold hells, some are hot hells, the deepest and largest one being Maha-Raurava—that means "Great Howling Hell"—you may remember your last stay there—I do!

Some people take you as a bad guy if you say that, but we all have a long, long past, and as Srila Prabhupada says, "Just because I am a very nice man in this life does not mean that I have always been a very nice man." Srila Prabhupada, thugh, is an eternal associate of the Lord and has never been captured by Maya. Krishnadas Kaviraj very famously considered himself lower than a worm in stool, but that does not mean that he was. That sort of extreme humility is a mystery of the devotional school—the closer we approach the Transcendence, the meaner we think of ourselves. That means I must be far away, because so far I have a pretty good opinion of myself.

We have been discussing the classical Vedic Universe of 7 worlds upwatd amf 7 downward, plus seven hells, but Guru Nanak says there are unlimited worlds in all directions. Prabhupada, most of us know, says we never went to the Moon—because he relies on the Vedic descriptions which say that there are forests there and a civilization more advanced than our own—the astronauts must have landed on Rahu instead—and indeed, that's what everyone really thinks, and why there was disappointment at the Moon landing instead of happiness. But this is what happens when you take the Vedas literally—or the Bible!

I follow the school that says nothing in the Vedas pertains to the physical world—if there is a tree in the Vedas, it is a meditsational tree, not a tree in the world of botany and physics. I have had enough of literalism from my HS religion teachers—I don't need more from the Krishnas! The ways in which they are the same as Christian fundamentalists and the antagonism the fundamentalists feel towards them is sometimes comical, although also sad.

One tends to get stuck on the Path—into stereotyped beliefs and speculations, ritualized habits, and so many things. In real spiritual life, our conception grows daily and changes—it is a living thing. It is one of the duties of the One Guru to unstuck us in this circumstance. Thus he appears in many Forms—"the real disciple must be able to recognize his Guru in a different Form". So long I took him as the Prabhupada Form, but he has appeared to me an Nanak, and my understanding has changed in reading the Japuji. Actually, I have five Gurus—Jesus, Prabhupada, Srila Sridhar Dev, Srila Bhaktisiddhanta Saraswati, and Nanak Dev. And the Five are One Guru, like the Pancha Tattva.

This is a journey of faith, the "evidence of things unseen", an essential spiritual substance with a tangible place in our sadhana—we may travel the Vedic Universes and see the Ocean of Milk and the Ocean of Sugarcane Juice, and many other places—the Guru will take us there—we can travel to all the pilgrimage places, Hardwar, Varinasi, the temple in Kholerganj, Vrndaban, etc. but it is just the "disease of the mind"—the pilgrimage sites will come to us if we desire rightly, and we can live at one while at home. When I was a homeless yogi living in the Oakland Rose Garden, I called it amritsar, because that's what it looked like to me. Years later, I saw a poster of amritsar and it is exactly as I thought1 And I have never been there this life. So pilgrimage places are moveable—Golok itself can come to you!

PART TWO

Life, Light, Love and Liberty

LLLL is an old, Gnostic signature. The Gnostics, or Jnanis, considered that God is Light—there are some Vedic statements to this effect, also. The Gnostics appear to have taken over a lot of ideas from the Indian monists—or developed along the same lines independently. Some Gnostics adopted Christianity, wrote their own Gospels, and proceeded on; others stayed pagan. LLLL is also a signature for some modern occult and Rosicrucian groups. One fellow I met at a seminary on Holy Hill was completely incredulous that anyone would want to go to a Gnostic Mass. "What's the good that?" he asked. "What's the good of anything religious?" I responded. "the Gnostics are not extinct and they have as much to offer today as they ever did back when they nearly overtook Christianity."

Personally, I do not care much for Gnosticism or its modern eruptions like Rev Moon, or the New Age in general, but I respect it as part of the occult tradition. There are also new, Christian Gnostic groups who promote the Gospel of Thomas, although they don't so much like the Gospel of Judas, a similar Gnostic production. I am not a Jnani—I am a Bhakta.

1. The Charter of Human Liberty

Liber AL vel Legis is presented as such in certain Orders—it confers not only liberty here in the land of liberty, but in the spiritual world—if you know how to read it!

I don't read that Book any more—"it is best to burn it after the first reading" (first commentary)

Ra-Hoor, for instance, describes himself as "in a secret fourfold work, the blasphemy against all the gods of men." Crowley says, "the secret fourfold word is Do what thou wilt."

And yet here I am, a Krishna bhakta and respector of all demigods.

How do you adjust that? The "gods of men" are all the things people worship instead of God: Money, cars, position and prestige, movie stars, music personalities etc. which fill their hearts instead of the One without a Second.

For a people that love to talk about freedom, we sure are a nation of busybodies. Everybody wants to mind everybody elses business about smoking (whether tobacco or marijuana), for instance, and then talk about how free we are in America! Helmets, seat belts, no-hands cell devices and child seats are momre of the same! Of course, this is the land of the dollar and money confers many freedoms—so we can enjoy the freedoms that money can buy—staying in nice hotels, driving a big car, wearing $800 suits, etc.—but you can do that in Singapore, also, or anywhere in the world. My point is that real Liberty is not conferred by a Bill of Rights, and can be pracriced anywhere, regardless of the local system of government—it is called Mastery of Self, ansd like St Paul says, "against this there is no law." It seems the mystically inclined mostly just want a peaceful status quo in which to seek moksha, or nirvana, or love of God—not revolutions, rebellions. Shootings, bombings, and mayhem—they would rather not have that. It has been said that "Democracy is the greatest system for bamboozling the people ever invented." And it seems to be true—in the democracies there are fewer revolts, revolutions, bombing, etc. than in less democratic countries—the people are fooled into believeing in the "representstive" system, and thus don't take things into their own hands, even when they should.

2. Gayattri Devi—Divine Light

Gayatri is a meditation on Light. It is probably the most important meditation in India, next to the chanting of the Holy Names. One receives the sacred thread in order to chant it—the thread represents sense-restraint, and to wear it, one takes a vow of celibacy towards all women/men except one's spouse. For many centuries, women were not offered gayatri—in modern times, they often are—only they wear the thread wrapped around their wrists instead of across the chest.

Actually, there are many gayatris; it is a poetic metre—there is one for Agni, one for Ganesha, and so forth, but I speak here of Brahma-Gayatri, the main one. Many powers and the development of many virtues are promised to one who 3x daily recites gayatri at sunrise, noon, and evening. Gayatri can be received from anyone who possesses it, or you can make your own link.

You can make your own thread, too—it should have three cotton strands and stretch across the chest 3x. There is a special knot holding it together called the brahma knot—but you should tie it according to your own ingenium—Krishna knows what you mean!

The link, or Magickal Link, Crowley says, is the "most important topic in all Magick." Everything is within us and outside of us—Sun. Moon, Golok, pilgrimage places, God—all within, but how to access them? Generally one receives his own wealth from another—but if this Is not possible, we understand that God is the Initiator. Get a picture of your favorite acarya, and he will show you what to do. If you do not have a favorite acarya, or have not accepted a Guru, you may take Dattatreya as your Guru for this rite: his mantra is OM DRAM DATTAYA NAMAHA. The Guru is not bound by time or space, or timespace, as it is called these days.

This is the mantra: OM BHUR OM BHUVAHA OM SWAHA OM MAHA OM JANAHA OM TAPAHA OM SATYAM OM TAT SAVITUR VARENYAM BHARGO DEVASYA DHIMAHI DHIYO YO NAHA PRACHODAYAT

The essential meaning is, "I meditate on that beautiful Savitri, Who lights all the Worlds—may She enlighten our minds."

The short version is : OM BHUR BHUVAHA SWAHA TAT SAVITUR VARENYAM BHARGO DEVASYA DHIMAHI DHIYO YO NAHA PRACHODAYAT

3. The oldest one

When the profane discuss different religions, the question always arises—which is the oldest one? Oldest is not necessarily better, is my principle—by one calculation, the oldest are from before the Flood—but people were drowned for practicing those religions so it may not be a good idea to take any of them up. According to another calculation, religion began from Prophet Adam, the first person—there never was a time when it was not, and monotheism did not develop as part of a process of evolution because religion is not invented.

The Vedas, we are assured, emanate from the breathing of the Supreme Personality of Godhead Maha-Vishnu—therefore they are infinite—no one could write them all down, not even Vedavyas, who could do the impossible. There is no end of variegated Vedas—but academic scholarship considers the Rg Veda the oldest scripture. There is sure to be an argument about the Bible and Judaism, however, if one asserts this.

The real religion is ever-existing—mankind is much older than the proponents of evolution think—and it has always had its adherents, sometimes few, sometimes many, and it is not tied to the doctrines of any special book, although its followers generally follow a book—

We must catch up the thread of the conception God and Guru bring to us and follow that, and not be poisoned by being caught in dogmas, doctrines, or too much rule-keeping—God is found through love, not rules.

"Which is the oldest' is therefore not a very intelligent question—"Which is the best for me?" is the cardinal question, if we admit Truth in all of them. Some say Jesus is the only way; that is true in the sense that He is the Christ, a special conception of God He had, and which is in Tiphareth, indicating that this conception is all Beauty, which is the attracting force, Christ-Krishna, but you can have a relationship with Jesus without becoming a Christian! It is simply unnecessary to accept all this Bible stuff, the Bible being a fifth-rate scripture, written for a primitive and barbaric people that could not understand very much about God. And if we are Jews, or something like it, we can accept Christ-principle and neglect the man Jesus.

So "the oldest" has not much to do with anything—different types of ideas are enshrined in different forms. The Hare Krishna religion has a long, long pedigree stretching way back here and on other planets—but its modern form was originated by Chaitanya Mahaprabhu in the 15th century—the form of congregational and street chanting, when by His Will there was a great revival of bhakti all over India. Bhakti actually means love with service, and it Is not intended for demigods—it is meant for the Supreme only. The Sikhs also were one of the groups that originated in this period—they are also bhaktas, and Guru Nanak met with Mahaprabhu on one occasion.

As I have said, I am a bit of a Sikh, and, understanding my mentality, Prabhupada and I had a talk. We discussed my writings, my concerns and realizations, and the fact that I could never be completely comfortable preaching the type of sexism that he has so enthusiastically expoused. I think what's sauce for the goose is sauce for the gander—for every sexist statement made and women, there is a corresponding one about men. I don't go for all this cutting down of women found in the branches of Vedas we know, even if it is true.

Anyhow, Guru Nanak didn't go for all this sexism, so Prabhupada sent me there. There are other reasons also—I am into a lot of different stuff, and so is Guru Nanak—I like Islam as well as KC—so does he—and there is a connection from previous lives as well.

Actually, HDG Paramahamsa Parivrjakacarya Bhaktisiddhanta Saraswati Goswami Prabhupada sent me here to America, and my process of realization has been to find out why! It wasn't to "help" Srila Prabhupada, the self-sufficient acarya—I missed out on most of that—but to study his translations and purports, practice his philosophy, master it, then add Thelema, Sikhism, Christian Science, Quabalah, and so many things, then write and publish about it. Prabhupada himself is not much interested in really understanding any other religion—his remarks about Christianity are frequently way off the mark—he said, for instance, that a "Christian is one who keeps all ten commandmets". No, Prabhupada, that's a Jew. A Christian is one who loves God with all his heart and his neighbor as himself—that's the commandment of Christ—the ten commandments were given by Moses, not Jesus.

4. Pranayama—Life

Prana means life—life energy—it circulates in the body, and concentrates at certain places, where chakras have been placed by meditators—the chakras are not real. It is a subject of major interest in traditional Oriental medicine—the Chinese call it chi, the Japanese ki. Pranayama means regulating this prana. This is generally done through the breath—pranayama can be very potent, as those who have practiced it know. The prana also can be regulated by mantra, like the Maha-mantra. The essential idea is, I am not the body, subtle body, desire body, etheric body, but I am senior to them all—"the devotional attitude is the spiritual body". I postulated all these things because I desired to enjoy matter, where there is no enjoyment, because there is no endurance—you may say, temporary pleasure is better than none at all—but the pleasure available in the worlds of Spirit is millions of times greater than the puny pleasures of sex and lordship we can enjoy here—the trouble is that they are mutually exclusive—the human form of life is declared the best form for spiritual realization—but it is, according to Srimad Bhagavatam, the best form for sex life, too—Choose ye well!

There are innumerable nadis (lit.rivers) in the subtle body, but there are three main conduits—ida, pinglala, and susumna. These correspond to the Three Pillars of the Tree—Severity, Mercy, and Mildness. Kundalini is like the Lightning Flash. When Kundalini meets Her Lord above the head, a nectar is released from the area of the brahma-randhra, or "hole at the top of the head" (through which yogis depart this world), which flows downward, drenching the chakras and conferring bliss, immortality, and mystic powers, which the Magician craves.

This brings up the whole topic of "awakening Kundalini". If you have read this far, your kundalini is already awakened. She is the Energy of Devotion, an aspect of Radharani, and She follows the Holy Names.

There are bijas (seed syllables) and mantras connected with each of the chakras—by, for instance, reciting the mantra to the svadhithana chakra, one can get a whole lot of girls! But the yogi doesn't want all these girls—so what is supposed to be the use of that? The maha-mantra, on the other hand, controls all the chakras.

The trouble is, hardly anyone wants to accept that the process of chanting Hare Krishna and taking spiritual food can accomplish all these difficult yoga proposals by itself—the garden variety seeker wants something difficult an dphysically challenging, full of secrets and arcane mysteries—some big deal yoga.

But all secrets and mysteries are contained in the Name of Krishna—it wll take you within, where you will learn them all. There is no harm in adding things, but the Name is the Supreme Secret of anything.

5. The Current of Pure Devotion

Axioms:

1. God Exists
2. He is very kind, and very beautiful
3. 3. He appears in various Forms
4. He is attained by Guru's grace
5. God is always the Guru, but the Guru is not always God without distinction.
6. God has many Names, according to His attributes and activities.
7. All of those Names are non-different from Him.
8. To chant these Names is to put an end to sorrow.
9. His personal, main Name is Krishna, or Krishan
10. By worshipping and remembering Him, we may attain anything or everything, according to our desire.
11. But we do not ask for anything in exchange for our devotion but more devotion.

A Current means like in the ocean—a powerful stream of awareness that we can get caught up in—Lord Chaitanya created a huge tsunami in the Ocean of Bliss and it inundated all of India even the Western countries had an uprising of bhakti and mysticism—it is still flowing and inundating the entire world with Krishna prem. This Current is not a dead current—it has evolved in the past and it continues to live and evolve, although the basics remain the same.

Bhaktivinode Thakur captured the idea by reading books—he reformed the Gaudiya Tradition, which had become a religion of low-life people. Like Varaha, Who lifted the Earth out of the waters, he lifted Lord Chaitanya's divine ideas up again. We also can capture the idea by reading Prabhupada's books.

Pure devotion means not to ask anything, and whatever dealings the Guru has with one, to accept them as being for one's best good. It means unattached to fruitive activities, one does not engage in mental speculation—that type of speculation that leads to the monistic conclusion. One has no other goal but the satisfaction of Guru and Krishna—as Lord Chaitanya says, "even if Krishna neglects me, He is my Lord, unconditionally.".

This ideal was not known to the developers of Quabalah—that is why there is so much magic in the Quabalistic tradition—if God is unapproachable, we turn to magic. The Krishna view is that god has His Pastimes in the supernal world—He "has a life" with the Gopis and his friends, and sometimes that life comes here, just to show us. The Quabalistic God, contrariwise, is like the big Eye in the pyramid on dollar bills—He watches us, day and night, rewarding some and punishing others, and that is all He does. There is the tradition of the Beard of the Ancient of Days, which comes down into the Sephiroth, and much speculation ensued about the meaning of the different parts of the beard. Krishna has no beard. He is eternally a youth of 16. Sanatan Goswami was a Magician—he had a touchstone which he kept near the garbage, and he made amulets for people—but he was also a pure devotee.

6. Meat or Vegetables?

We know that Adonai allowed the Jews to eat meat, because they craved it. Kosher gradually developed, and just like the Muslims, the Jews took to the idea that it was ungodly to be a vegetarian. That is one of the differences between the Abrahamic religions and those coming out of India—it is a major difference. Manu says vegetarianism is meritorious, but not required.

Srila Prabhupada takes a special tack in this debate. He says bhoga, food, should be offered to Krishna before it is taken. Krishna happens to be a vegetarian, so meat cannot be offered to Him. And since we only eat offered food, we are vegetarians, also. His evidence for this is a verse in the Bhagavad Gita where Krishna says, "if you offer me a flower, fruit, leaf or water, I will accept it." Vegetarian foods being transformations of fruits, flowers, etc. Krishna will accept them. But I take note of the following verse: "Whatever you eat, do it as an offering unto Me." "Whatever" would mean all sorts of meat—so I am not convinced of this—I will have to ask Srila Prabhupada. Of course, it is good not to be cruel—we don't see the cruelty in the slaughterhouses, but it is there. And especially in the Krishna movement, there are all kinds of delicious vegetarian preps, proving that there is no necessity for meat-eating. Manu said it was very meritorious to be a vegetarian, but not required—apparently even in India in his time, many people ate meat. Prabhupada always brings up that goat—"if you want to eat meat, you can sacrifice a goat to the Goddess Kali". But I would imagine one would get very tired of tough, stringy goat meat all the time, and want to eat a chicken, or a lamb, or something, or fish. Nothing but goat is a terrible idea. Similarly, I have argued with people over Van Camp's Pork and beans—there is only a tiny bit of pork in pork and beans, so to me it does not matter—to some people a piece of pork half an inch long is a big, big deal. I am told that if I want to follow Srila Prabhupada, I cannot eat that tiny piece of pork. I don't thik that that is true, but I usually lose these arguments. But I am not a "food religionist", strictly keeping vegetarianism, or kosher, or halal, or whatever it might be—I think Prabhupada has his rules, and I have mine. My rule is from the Fama Fraternatis: "Eat the common food of the country you are in"—the rules for the Rosicrucian Brotherhood—not AMORC, in San Jose, but the real Rosicrucian Brotherhood., the invisible one.

7. Tree of Household Life

Jews and Sikhs are householders—so are Muslims. None of those religions has much scope for single, celibate men or women in any sort of monastic existence. Hinduism and Buddhism do—in the strict monistic sects, if one is serious he becomes a monk, and this estate is quite respectable. Guru Nanak did not like the ascetics, wandering around begging—neither did Prabhupada. St Paul approved very much of the single, celibate estate because he was that way, also. Of course, there are single people even in a householder culture, due to death and sometimes divorce, and religious cultures frown on remarriage so there are many single widows and abandoned wives.

The question is always asked of swamis—can a householder attain God? The answer is yes. But it is more difficult, with so many affairs to attend to, worries about money, quarrels with one's spouse, and so forth. It is much more difficult, like Jesus said about rich men—and the disciples were astonished—I don't see why! Money tends to make you forget God, and the more you have, the easier it is to forget—"Who, then, can be saved'?"—the people who don't forget! That is one of the easiest statements of Jesus to understand—He said some very enigmatic things.

The names of the Sephiroth are:

1. Crown
2. Wisdom
3. Undererstanding
4. Mercy
5. Severity
6. Beauty
7. Victory
8. Splendour

9. Foundation
10. Kingdom

The colors (King scale) are:

1. brilliance
2. pure soft blue
3. crimson
4. deep violet
5. orange
6. clear pink rose
7. amber
8. violet purple
9. indigo
10. yellow.

One of the troubles with asceticism is that, as Crowley says, it invariably leads to spiritual pride. Jaggadananda Pandit, an associate of Lord Chaitanya's, instructs that it is very dangerous to discriminate between a householder devotee and a renounced devotee—many offenses come from this. Of course, "householder", "grhastha" had a different meaning in medieval India than it does today. This "household" was an ashram, devoted to serving the wandering sadhus with no other means of support and to pereforming sacrifices and raising children who would transcend birth and death by good instruction of their parents. Today's householder is not a grhastha, he is a grhamedhi, one whose center of interest is his family and who has no higher vision of life. "Do not produce," Srila Prabhupada says, "children like the cats and dogs". the idea is if you can free your children from birth and death, you can have a hundred of them. If you can't, you need not have any.

Household life was envisioned by the sages as a sacrificial fire—marriage itself is considered a sacrifice. There are four Orders of life in the old civilization, and three of them do not have any money—the householder must support the other three Orders of society, and therefore his main task ion life is earning money.

The sexual restrictions are, I think, too severe in Prabhupada's ISKCON. He allows sex only for procreation, never for fun, and I thik this is because he does not like girls and does not want to give them any satisfaction at all. He did not like his mother and he did not like his wife—and these are the most important women in a man's life. Being connected with the Vedic tradition which is itself very sexist, gae him lots of scope for his own sexism. Myself, I think what's good for the goose is good for the gander—for every negative statement made about women, there is a corresponding statement about the faults of men. The Vedas seem one-sided in this regard—but we haven't seen all of them. Prophet Muhammed (PBUH) is much more liberal I n this regard, although if you try to tell the devotees that, they will begin blaspheming him. So was Guru Nanak—the only restriction they made was that one be married. Besides that, they stay out of your bedroom, which I think is the best thing—neither Pope nor Parbhupada in bed with me1

8. The Art of Chanting on Beads

The fact is, chanting 16 or more rounds a day is a serious yogic accomplishment—it is not at all easy—and there are many bebefits like control of the mind, control of the prana, and so forth. It is not at all easy. This is because the Names are bitter as poison to us—we hate them—e do not at all like the Lord's Names—it is like one who has jaundice finds sugar candy very bitter—but if he continues to eat sugar candy, he recovers and can taste the sweetness (Rupa Goswami) similarly, if we can force ourselves to chant, we will recover a little taste for the holy Names eventually, and then the task will become pleasurable more and more, provided we avoid offenses, which are ten. Chanting with offenses is called nam aparadh, and it won't get us very far on the Tree of Bhakti—I have arranged the offenses on the Tree so we can see how they block the current of bhakti. Next is the stage of clearing offenses—namabhasa—when offenses have stopped but one is not yet on the purest platform—"by namabhas one can get liberation" This is all discussed in a book by Bhaktivinode Thakur called the Holy Name—the final stage is shuddha nam, or pure name, and it is beyond liberation or any conceivable goal. The ten offenses are:

1. To defame the Vaishnavas
2. to consider that other living entities, such as the demigods, are independent of Krishna
3. to neglect the Guru;s commands
4. to disrespect the scriptures
5. to interpret te meaning of the Name or to think that the glories of the Name are mere eulogy
6. to commit sins on the strength of shanting

7. to give the Name to the unfaithful who are not ready
8. to accept
9. to consider the Namd eual to pious works
10. to be inattentive while chanting the holy Names
11. to continue to be attached to the material world despite having heard so many instructions on this matter

of these, number 6 is the worst and most serious of offenses—but again, it depends on the development of one's faith—what IS sin, and are these things really sins, and so forth—not that we are just given a list of things not to do and if we do them that is "sin"—a little more realized and sophisticated than that, I think!

For those who believe that the "word of sin is restriction" it is certainly possible to sin against oneself, and that is probably the worst sin—as the anonymous sadhu says, 'If you think about sin all the time, you will become a sinner." Do Not restrict the flow of Love from the highest—you can choke it up by your belief in your own mortality—this is called free will. Anything that asserts that mortality against your immortality as Spirit is sin. Sin means forgetfulness of God, and who remembers? "He Himself remembers Himself"—that is the process.

On the Tree there are Mercy and Severity but no Evil, I should point out—some of God's aspects are shewn, He has 64 qualities, and the Tree may be His Countenance, or it may be His backside, which is all He was willing to show to the Prophet. In some traditions, there is an averse Tree called the Qlipoth, in which Kether is replaced by by a duality called the Clangers. This averse Tree fascinates some people, but we are not going to go into it.

'Disobeying the Guru's commands" primarily means the four regulative principles—they correspond to the four legs of the Bull of Dharma, or religion, as well as to other groups of four, like the four Quabalistic Worlds and the four letters of Tetragrammaton, thus:

No illicit sex—cleanliness—assiah—Y
No meat-eating—mercy—briah—H
No intoxication-truthfulness—yetzirah—V
No gambling—austerity—atziluth—H

The principles themselves are an Incarnation of Guru, being made of his words. Their influence is very broad, and keeping them leads to what Buddhists call "the awesome demeanor" which is so important in preaching.

"The first preaching is to oneself"—to try to understand enough to say something in regards to God, Nature, and we living entities pr jiva souls. One might only understand that Krishna is the SPG—that would be enough to begin the process—one will always meet people who say He is not—but for me, the "true is the useful, the useful is the true" and it is highly useful in one's life to understand Who is the SPG, especially if this information is coming in parampara from a real Source. The parampara is a little tricky to put on the Tree—I would be inclined to put Prabhupada in Malkuth, as the current link, Bhaktisiddhanta in Tiphareth, and Bhaktivinode in Keter, although there are many other arrangements, as there are far more acaryas than there are Sephiroth.

The Sikh Gurus, however happen to be 10;

1. Nanak
2. Angad
3. Amar Das
4. Ram Das
5. Arjan
6. Hargobind
7. Har Rai
8. Har Krishan
9. Teg Bahadur
10. Govind Singh

The 11th Guru is the book, Guru Granth Sahib, which should also fit unto Malkuth. Take note Har Krishan is simply Hare Krishna, so when we are chanting we are invoking Guru Har Krishan also. I mention the Sikh Gurus because Sikhism is a very progressive part of the religions of India—based on Naam, or Holy Names, once again it has a long history of Adepts who practiced the Naam before Nank started putting together what were tp be Sikhs—servants of the Guru.

Discipleship in the New Aeon

We are the Guru's servants, not his slaves; it is, perhaps, better to be his slaves—first slave, and learn the Vedas, learn about God and Nature and the position of the living entity, and when one has learned—servant. If I want to eat a hamburger—Do what thou wilt shall be the whole of the law! If I want some sex—Do what thou wilt shall be the whole of the law!—there is Liberty in the guru-disciple relationship, just like with Krishna—once surrendered, you are His regardless of what you do! Similarly, the Guru takes charge of you forever—the relationship is eternal—this news should fill us witrh confidence and peace, because we are prone to satisfy the senses.

One is always warned, however, not to trouble the Guru by committing any kind of sinful activity—he takes one's karma on himself, so he will have to suffer for that hamburger1 But how much suffering could there be for a lousy hamburger? Feeding self is not a sinful activity—besides "If he be a King thou canst not hurt him".

So what, then, is a disciple?

A disciple is a student who has accepted a discipline. His duties are to hear, inquire, and serve. "Tad viddhi pranipatena, pariprasnena sevaya; upadeksyanti te jnanam, jnaninas tattva-darshinaha" 'Just try to learn the truth by approaching a spiritual master. Inquire from him submissively and render service unto him. The self-ralizaed soul can impart knowledge unto you because he has seen the truth".

No one can get free from the influence of material nature by personal attempts; one must accept a bonafide spiritual master and act under his direction. (the spiritual master and the disciple, p.42)

The bonafide spiritual master need not be embodied—one may be a disciple of Christ, or Nanak, or the current or any of the past Gaudiya acaryas (not neglecting the current link), or the spiritual master may not even be a person—he may be the shabad, the Divine Sound, or the Light, or Chaos and Old Night in some cases—the Tree itself is a form of Guru, if you can work it. And despite Jesus' saying, one may serve two Masters or more, when they are all the One Guru; Dattatreya had 24 gurus, I have five.

The institution of brahmacarya has been revived by Srila Prabhupada, and I recommend trying it out, to learn what real discipleship is all about. Commonly, brahmacarya means celibacy, but it means a great deal more than that—it means "coursing in the Brahman".

Some of the qualities of a disciple are:

Trust
Enthusiasm
Cleanliness
Humility

Energy
Attentiveness
Love for the master
A great desire to please the master

Initiation means to go in, to get inside. It is the "process by which transcendental knowledge is imparted to the disciple". Plainly, this constitutes many things, not just a ceremony which is called "initiation". It may come piecemeal, or without any ceremony. Since God, Krishna, is infinite, there is no end to the process of initiation—one is ever being initiated into new aspects of the Lord.

The duties of the Guru are: To teach, to protect, and to chastise. Srila Prabhupada is teaching the whole Universe through his books—no other so-called spiritual group has books like his—each book is an Incarnation of Mercy, and these books will eventually inundate the World and all other planets with the current of Krishna prema.

The Guru protects us by his mystic powers. And he instructs us not to do things which will cause us trouble later, and thus protects us from karmic entanglements. Because he is as good as God, he can hear our prayers.

And he chastises his disciples to teach them—sometimes this chastisement is very embarrassing. Without chastisement there is no learning—the disciples are privileged to undergo the chastisement of the equal-minded Guru.

There is a great relationship of love and friendship between the disciple and the Guru, mixed with awe. It is similar to the relationships which abound in the Vaikuntha planets with Vishnu. And as the Law penetrates one's being, one more and more asserts his liberty in this relationship—as the Prophet (PBUH) declares, "there is no compulsion in religion". Everything is a free relationship with the Master, although not necessarily with the Temple authorities—I recall a time when I was staying at the Chicago Temple and I had a sleep disorder—I couldn't do anything other than sleep all the time. They took it is a moral fault, and were very displeased with me and forced me to work—I was not appreciative of that.

Fighting on the Battlefield

"The Lord Your God is a Mighty Man of War."—Bible

"senayor ubhayor madhye"
in the midst of both armies

Life is a battle—and we are all, apparently, cut down on the battlefield by eternal Time. This is the apparency. Actually, we are eternal—only the body, made of matter, "dies"—as Soul we have neither birth nor death nor sin nor disease—we think we do, and so it becomes. So what is the battle about? For some, it is Truth verses error, Love verses lust; for some it is simply to get the most toys—so we all understand this analogy.

But the Kuruksetra battle is not an analogy—it was an actual fight, and Krishna and Arjuna actually stood on the chariot in the midst of the two armies, and Krishna actually spoke the Bhagavad Gita there, and displayed his terrible Universal Form which is so attractive to the less intelligent in devotional service.

Srila Prabhupada was asked about Gopi pastimes once, and whether we should try to enter those pastimes—he replied, "We are not in the Gopi pastimes—we are in the Kuruksetra pastimes, fighting on the battlefield".

In one sense, all of Krishna's pastimes are of one and the same quality—you can place the pastimes on the Tree—the various demons slain, kings defeated, princesses married—all this information is in the 10th Canto of SB (Krishna book) and I will not essay to do it for you—better you apprehend these things for yourself.

So Bhagavad Gita means the Song of Bhagavan, and Bhagavan is a fancy Sanskrit name for the All-Opulent. Some people take this designation for themselves, but really I belongs only to God and those, like Narada, who are practically God. It and its message are on the so-called Ray of Devotion—it is said that there are other Rays, and other approaches to the Supreme, but the Gita says no, bhaktya mam abhijanati—by bhakti alone is the Supreme Absolute Truth approached—those who try to approach by speculation also reach Him—after many, many births.

Helping humanity is a noble cause—helping all sentient beings is a nobler cause—but serving the SPG is like watering the root of the Tree—automatically all these other services are done, because He is the Soul of all creatures.

"Hallowed by Thy Name"—the most mystical method of service to the SPG is to take His Name—this is a very mystical thing—He has many Names, like Rama and govinda, but the True Name is not the physical sounds of these Names—it is received by surrender to the Master who possesses it, and there is no timetable for that.

The Master B.R.Sridhar gave me the Name; it was confirmed and protected by many other Masters—and it must be cultivated constantly (or regularly—abhyasa yoga). By ghe grace of the Name, or Naam, we obtain everything. (Guru Nanak)

"God and His Name are the same."

"Work out your own salvation with diligence—decay in inherent in all component things."—Sakyamuni Buddha

I can only give the letters;:

Hare Krishna Hare Krishna Krihsna Krishna Hare Hare/
Hare Rama Hare Rama Rama Rama Hare Hare

The Great Work—and the learning of one's True Will

Sanatan Goswami, one of the individuals fully in the current of rasa, or mellows with the SPG and His Consort, was an occultist. He possessed the Stone of the Philosophers, generally considered the culmination of the Great Work, but he kept it near the garbage. One day an individual came to him, seeking the "best thing"—sizing him up, Sanatan gave him the Stone. The man was happy for a while, but he began considering, why did Goswamiji keep the Stone near the garbage, if it was the best thing? So he returned, and Sanatan told him to go and throw the Stone in the Ganges, and he would give him the best thing. The man did so, and Sanatan initiated him into the chanting of Hare Krishna.

So the Great Work may vary as to what one finds important at the time—certainly self-realization is a great work, whether carried out by yoga or by magick—I consider that having come to the Vaishnava Guru and accepted him, and having been accepted, that I have accomplished the Great Work—there is, as Krishna says in Gita, "nothing left to know". The rest is simply details.

This attainment includes the knowledge of one's True Will—True Will is only found when one eliminates all mortal will by serving the spiritual master. Then it becomes very clear, what this particular embodiment is meant for—the will-er being Spirit, the Immortal One Who knows all about whither we have come and why. That is called Ruach in Quabalah.

True Will is a metaphysical entity—it cannot be mutable, or born of matter. Although it manifests in this world, it is yet beyond the restrictions of this world, mainly the Three Modes of Material Nature.

These Modes are: Goodness, Passion, and Ignorance. Everything here is made up of some combination of these Three Modes. Goodness binds to happiness—Passion to intense endeavor, and Ignorance to madness—and sleep. There is a painting showing a man tied to puppet strings, and the strings are being pulled by the personified Modes—such is our situation—"One who sees that no other actor is a work here than the Three Modes, and sees the Lord Who is above them, becomes free from them." This method is explained by Srila Bhaktivinode Thaker thus:

"The path for attaining rasa is as follows: The jiva who has attained faith in bhakti receives a pure Vaishnava Guru. That Guru gives the maha-mantra. The jiva will take the mantra, either in the form of remembering the mantra or in the form of chanting the name, regulating himself by counting on tulasi beads. Gradually, as his craving for the name increases, he will increase to 3x64 rounds. Of the two forms of taking the name, kirtana (chanting) is the more powerful, for in that process are combined chanting, hearing, and remembering, and by it the senses of the jiva dance in joy. By taking shelter of any of the nine processes of devotion, one progresses in devotion, but of the nine, chanting is best. Those who have an agttraction for Diety worship will reach perfection only by also engaging in hearing and chanting the name. But those who have exclusive attraction for the name need only engage in chanting, hearing and remembering the name." The Holy Name, by Bhaktivinode Thakur p.113

So we are all wandering through material existence—although we are spirit-soul we have it in our diseased consciousness to enjoy the non-existent—in this, our guide is our desires, and we enjoy and suffer in the land of exploitation.

Srila Sridhar says there are three planes—the plane of exploitation, the plane of renunciation, and the plane of dedication. When we are done with exploitation, we think of putting an end—Buddha and Jain come to help us in this endeavor—we fall into a deep, dreamless sleep called Samadhi—but this is not the positive life of the soul—that is found in the plane of dedication—complete dedication to the Center, the Living Lord—and this is the key to True Will, also.

When we speak of dedication we speak of the High Shakti, the High Isis; the Lower Isis is Nature—the High Isis is immortality—She is Rukmini, Krishna's chief queen, and more, Radharani, Krishna's girklfriend, Who is the bliss shakti, Hladini Shakti. This should not be misinterpreted, as a Frater did, who thought She was bliss for him—She is bliss for Krishna, not for us—we are Her servants—we get bliss through service, not trhugh eros.

"One who is devoted to the SPG can attain all of the benefits derived from other yogic processes, rituals, speculation, sacrifices, etc. That is the specific benediction of devotional service'.

BGAI 12:7, purport by Srila Prabhupada

This is a good reason to lay down your sword and pentacle, down by the riverside. We are looking for the one thing we can do, by which all our purposes can be served all at once. Otherwise we will do this ritual and that, this practice and that, and there is no end of it though Time itself stop. Gayatri, for instance, grants all virtues and wishes, and that practice is for life—not 30 days of Gayatri and 30 days of Resh, and two weeks of another thing—fact is, Crowley, who is the Master of so many Magicians, was a scatterbrain. Actually, I went from him to a Hindu scatterbrain—a religion in which it is very easy to be scatterbrained—what is they say, "the Magicians Will must be single" and with Christ, it is the eye which must be single. And what does it say on the dollar bill? E Pluribus Unum—the Many are One. Time to use the Sword to cut off unnecessary vocations, and realize self.

One is devoted because he has realized himself as soul—otherwise he cannot do it—devotion to the Supreme is not an ordinary state of consciousness—we can devote ourselves to so many things—to painting or writing or pursuing a particular girl—that does not count as devotion. Devotion to the Supreme means with love and service, consciousness to greater consciousness.

Wrapping Up

It has been said that it is the nature of the Magician to associate things, in longer and longer chains of associations; just so, it is the nature of the Yogi to eliminate thngs—"elimination and new acceptance". It is best to eliminate one's whole material consciousness and accept the shelter of Spirit—"Who can be a better Guide than Allah/"—and everyone makes links and, mostly, the links already exist—Keter is in Malkut. Lord Siva has a connection with matter—He is married to Parvati, Nature, and has given Her half His body—but Krishna, Panduranga, has no such connection—He is pure Spirit—He simply glanced over the material Nature and everything began—He has no responsibility for it, and although He sometimes intercedes on behalf of His devotee, he is not directly interested in affairs here—there is a verse in the Qu'ran: "When good comes to you, think that it is from Allah; when bad comes, thk that it is from yourself." Although not directly involved, Krishna is always blessing us with more good than we can accept.

"Material nature and the living entities should be understood to be beginningless. Their transformations and the modes of matter are products of material nature, Nature is said to be the cause of all material activities and effects, whereas the living entity is the cause of the various sufferings and enjoyments in this world.'

BGAI 13: 20-21

Nature is the womb of the living entities, and Krishna is the seed-giving Father. Father means we are all taken care of by Krishna and our needs are arranged for—it may not seem so all the time, but that is a defect in our vision.

Because matter cannot feel, it is said that the living entities, who have sentience, are the cause of suffering and enjoyment—swimming in the sea of material good and evil, so-called, and sometimes drowning.

In this sea, we grasp the Tree of Life—Malkut is not exactly in this manifestation—it is difficult to become situated—it is the Guru's Feet and They are not quite situated here—the paths lead upward, to the liberated state, and there are many lights and sounds and beings there; when we come to Tipareth we can rest—Tipareth is the Guru's Heart; from there, we can fly straight as an arrow to Keter, which is the Guru's Word—and Malkut is in Keter, so we have traveled far without taking a step.

It begins and ends, not with Energy but with the Energetic. That is the real story—the dream is of Nature—we are born, grow into youth, into manhood, age, old age, and the body falls away, to return to the Earth. The story of the unborn soul, however, is only gotten through scriptures like BGAI; I need not tell you the Bible is not the only scripture-class document in the Universe; it was written for a primitive and barbaric people that could not understand very much about God, and therefore, although "inspired", is a fifth rate scripture.

A final word about Reality—some say this world is unreal—some only say, temporary. Vedavyas, a great sage, calls it unreal in the first verse of his Srimad Bhagavatam. I agree entirely with Mary Baker Eddy about this—she says, "Spirit is true and real—and matter is Spirit's opposite". It is like eating an ice cream cone—the ice cream seems something very substantial, but when its eaten up, its gone, nothing, nada—this world is like that—Maya means that which is not—so many relationships, wife, kids, friends, big job—just like the ice cream cone. So lets not waste time with the impermanent and the non-existent, shall we? Lets find the One Reality which is Infinite and Eternal and make our home there with Him.

Om Tat Sat
Sri Clayton, Avadhuta